Whatever Happened to "Eureka"?

Nick Downes

RUTGERS UNIVERSITY PRESS New Brunswick, New Jersey

Whatever Happened to "Eureka"?
CARTOONS ON SCIENCE

Library of Congress Cataloging-in-Publication Data

Downes, Nick.
    Whatever happened to "Eureka"? : cartoons on science / by Nick
Downes.
      p.   cm.
   ISBN 0-8135-2146-7
    1. Science—Caricatures and cartoons.   2. American wit and hu-
mor, Pictorial.   I. Title.
NC1429.D585A4   1995
741.5'973—dc20   94-25454
                                            CIP
British Cataloging-in-Publication information available

Whatever Happened to "Eureka"?

"Everyone talks of the hazards of living in L.A., but the last earthquake caused a mudslide, which extinguished a wild brush fire that would have consumed our house, so I can't complain."

1

"Defying the law of gravity, Sarge."

5

"It'll never heal if you keep picking at it."

"Normally, of course, Edward wouldn't harm a fly."

NICK DOWNES

9

"Take all six, I'll throw in the giant asteroid."

"Professor! Another lost episode of 'The Honeymooners!'"

11

"I worry about the subliminal message."

"Entire galaxies are drifting away from one another, Helen. Why should you and I be any different?"

"I suppose what bothers me is that it's the one job where you can't say, 'Relax—it's not like you're performing brain surgery.'"

"Of course he was cut down in his prime. Don't you get it? We're *all* cut down in our prime."

"What are all these bills from Genentech Inc.?"

"It's fast, but I miss the chewing and swallowing."

ASTEROID IMPACTS HAVE CAUSED MAJOR CHANGES
THROUGH OUT OUR EARTH'S HISTORY.

EXAMPLE # 34

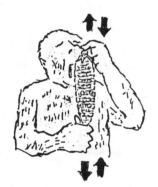

BEFORE IMPACT

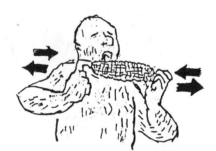

AFTER IMPACT

N. Downes

18

"The metallurgical, instrumental, elemental, microscopic, and laser light analyses say the butler did it."

"In your case, hindsight is 20–40."

21

"Have you noticed that no two snowflake researchers are alike?"

"We need to moth-ball plant 3, sir. We've got moths."

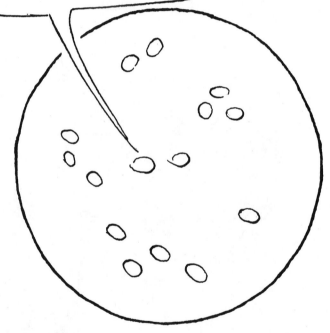

"We were a control group back in Behavioral Sciences 101, and we've been together ever since."

"One quintillion and three . . . One quintillion and four . . ."

"Whatever happened to 'Eureka?'"

"It would appear that your constant anxiety over health-care costs has resulted in a peptic ulcer."

"Relax, folks. You've treated me pretty good."

"None for Richard. Alcohol activates his reptilian brain."

"It was touch and go for a while. I'm lucky to be dead."

"Thousands of years ago, herbivorous animals wandered into these tar-pits, got stuck, and were followed in by predators—who also found themselves mired, eventually attracting paleontologists such as ourselves."

"They're natural enemies of roaches."

35

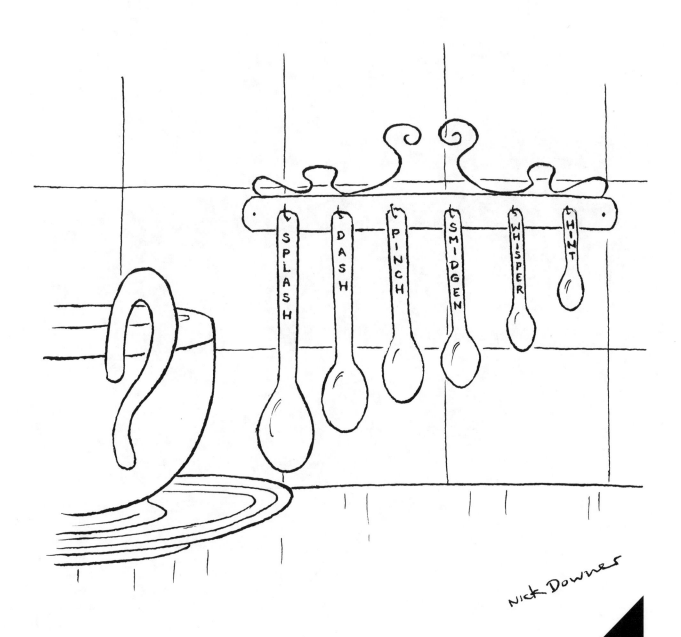

37

"'It is more important to have beauty in one's equations than to have them fit experiments.'—Paul Adrian Maurice Dirac."

"How soon can she begin tennis, Doctor?"

"And this crystal insures that any hypothesis I happen to be testing stands up to the deductive process."

"Oh, that's Edward and his fight or flight mechanism."

"It's a little late in the game to believe in sacred cows, Harry."

"I've been putting up with their insinuations that I'm a lousy math teacher for this many years!"

"The car pollutes the air, depletes the ozone, warms the planet, acidifies the water, and ruins human health. Face it, Dad—driving to work is a terrorist act."

47

"He and I go way back."

"Well, I first noticed the rash about 6 weeks ago, and, uh, look—are you sure you're a dermatologist?"

"Honey, I think you'd better put on some sunscreen."

"We'll be needing a lawyer."

"That's Dead-eye Dan, known far and wide for his fast gun, mean temper, and extra Y chromosome."

HERE LIES
JOHNOTHAN JONES
BORN   1911
STARTED OVER 1932
BEGAN ANEW 1946
STARTED FRESH 1958
TURNED OVER
A NEW LEAF   1967
BORN AGAIN   1975
DIED   1988

Nick Downes

53

"He's consistently right in proving his experiments wrong."

"Look—who cares what the nitrate level is?"

55

"Now, see? We never had those around here before this global warming thing."

"I paint what I smell."

"Empty-nest syndrome." Nick Downs

"It was so terribly sudden."

"He's hardly how I imagined him."

NICK DOWNES

"We can expect scattered showers tomorrow, with rain falling on the just, but not the unjust."

"And how long have you had this fear of spontaneous combustion?"

63

"We could convert to a computer-dating service."

"I'm putting you on a different brand of cigarettes."

"He may seem harmless now, but you haven't heard him sing 'Moonlight in Vermont.'"

"The obstetrician delivered you."

IMPERSONAL COMPUTER

"Speaking of liquid boosters, I see it's almost happy hour."

"There was a time when I thought the Earth revolved around *her*."

"Due to a butterfly flapping its wings someplace, yesterday's forecast was completely botched."

73

"You're carrying an awful lot of baggage from your childhood."

"Of course, within the species, there are those who are more endangered than others."

"I think maybe we represent spiraling health care costs."

"Hadn't we ought to be getting on with the case, Holmes, rather than burning insects?"

"We finally make it to the Yucatan and now this."

"Of course you're furious over the price of your medication,
Mrs. Grimwald—that's one of its side effects."

"Mrs. Pierce, your choice of this brand of spaghetti sauce contradicts your previous purchase history."

"What would be its overall effectiveness against, say, the Antichrist?"

BIOTECHNOLOGY

"Be my valentine?"

"This bread's been de-germinated, de-fibered, refined, processed, and bleached; on the other hand, it does have a picture of a farm on it."

"Oh, Ma'am—I wouldn't bother him while he's eating!"

SHOCK RADIO ASTRONOMER

87

"What a day. Two Jumbos and a Grade A Large."

"Now, of course, comes the awful wait for the other one to drop."

"I told you if we didn't tie him up he'd follow us."

"Your husband's cluster-transplant became a bit more far-reaching than we anticipated, Mrs. Sampson—we've replaced him with another man entirely."

"Somewhere along the line, his magnetic personality reversed fields."

NICK DOWNES

93

"HEAL!"

"He was a pioneer in his field, and he never lets you forget it."

"Now, Roger, there are so few places for a shaman left."

"It was either do my homework on dioxin-producing bleached paper, Mr. Dillard, or do my bit for the environment."

"And Bernie here makes book on when the big one will hit."

"There's a rather menacing cyborg from the future to see you, sir. Shall I tell him to come back, say, yesterday?"

"At least you cleared the spent-fuel pond."

"If symptoms persist, may God have mercy on your soul."

103

"Tit for tat, I suppose. He was a vegetarian."

"They say with high-definition TV we'll be able to tell one show from the next."

"Ever since he was implanted with human fetal tissue, he's been impossible."

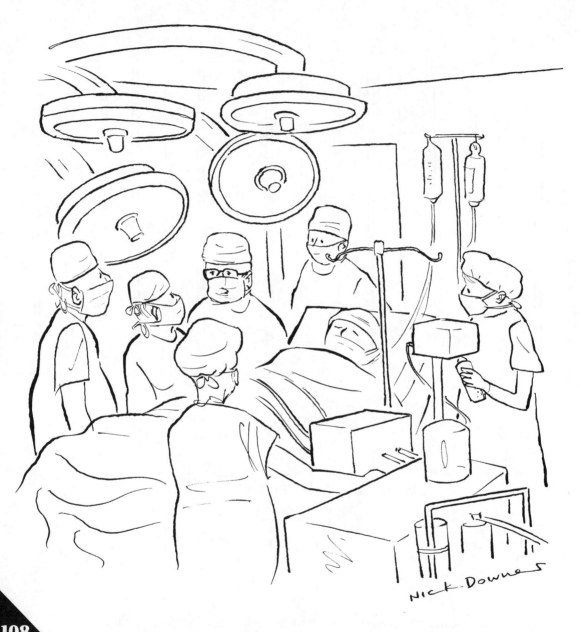

"As soon as the patient's insurance rep gets here we can begin."

"May I remind you that 99 percent of all species that ever existed are today extinct."

"Breathing. Most of us take it for granted until a problem develops."

"The universe is 10 billion years old. You don't think that calls for a little celebration?"

111

"Do you ever get the feeling we're not the only global marketplace in the universe?"

"It's been a wonderful whale-watch, but shouldn't we be getting back?"

"Hey! Where's my science project?"

"One can go through an entire medical career without encountering such a hangover."

117

"When I was your age, I had to walk 6 feet, change the channel by hand, and then walk the 6 feet back."

"No, I'm not a ghost, Mr. Scrooge, but I'm one highly agitated hologram!"

"I'm the store's marketing analyst. I come with the cart."

"I wonder if it's not too late to sue my obstetrician."

"I've been giving a lot of thought lately to modifying my behavior."

"He's just learned that by the time he's 18, he'll have seen only 52,000 murders on TV while his friend Billy, who's got cable, will have seen 72,000."

"That was one of the strongest solar flares in some time."

"As long as they limit it to plants, genetic engineering is fine with me."

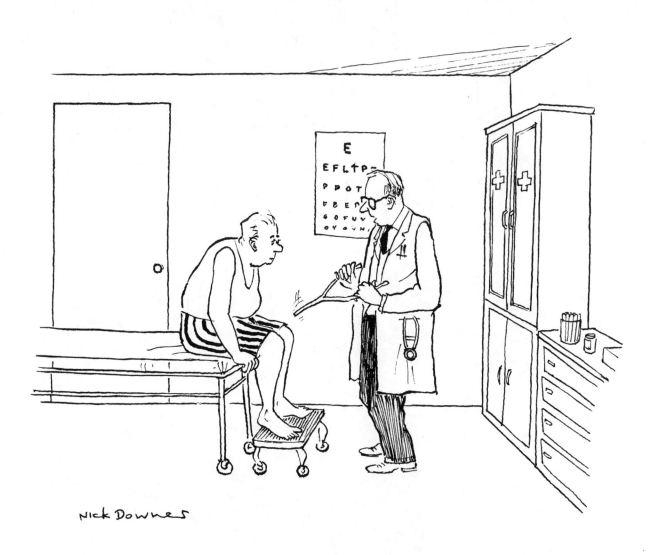

NICK DOWNES

"You've got water on the knee."

"One of the last living members of the Flat Earth Society, sir."

"Nice going, Griswald. You had to ask it who made God."

"How old are you?"

"Well, thanks. I find you rather Picasso-esque as well."

"It's George! He's collapsed under the weight of his own gravity!"

"And now, for my next compound . . ."

"It disturbs me that so many of today's young people go into medicine for the money. In my time, the chance to play God was enough."

135

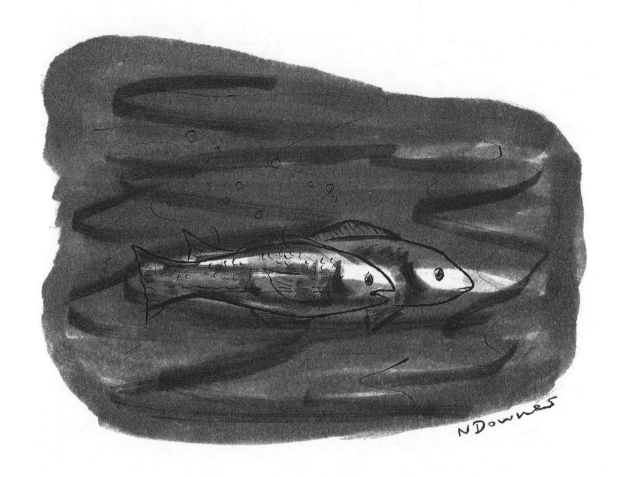

"Sure I was relieved. But then I started thinking, 'Why *did* he throw me back?'"

"I thought we were 80% water."

"Yes, Billy, but Mr. Phillips pushes legal drugs."

"I won't mince words, Mrs. Horton. Concerning your husband, we've had a negative patient-care outcome."

"And while 'Nature-Bran' may contain carcinogens, they're all natural carcinogens."

140

"Just browsing, thanks."

"Did you wash the lettuce?"

"This is an example of how various chemical emissions can combine in the air to create a health threat far greater than the sum of their parts."

"If you could blow that blues through a horn, you'd make a very comfortable living."

"Conversely, time slows down when you're *not* having fun."

"It may be unnecessary surgery as far as you're concerned, Mrs. Farris, but for me it's crucial."

147

"I'm sorry—I assumed you two were together."

"Oh, I see. I always thought 'under the counter' was a figure of speech."

"Expect showers today, with rain changing from black to sepia by mid-afternoon."

"Uppers or downers?"

"I've managed to discourage the medical-supply salesmen, Doctor, but there's a leech farmer out here I just can't shake."

"You'd think it would make him more sympathetic towards animal rights."

ABOUT THE CARTOONIST

Cartoons by Nick Downes have appeared in a wide variety of American and British magazines and newspapers, including *Punch, Saturday Evening Post, Science, Natural History, Philadelphia Inquirer, The Listener, The Spectator, Private Eye, Woman's World,* and *Amazing Stories.* His first collection of cartoons was *Big Science* (AAAS Press; also available from Rutgers University Press). Nick Downes lives in Brooklyn.